Leonardo's Dessert
No Pi

Herbert Wills III

National Council of
Teachers of Mathematics

Library of Congress Cataloging in Publication Data:

Wills, Herbert.
  Leonardo's dessert, no pi.

Bibliography: p.
1. Geometry—History.   2. Leonardo, da Vinci,
1452–1519.   I. Title.
QA443.5.W55   1984   516   84-27185
ISBN 0-87353-221-X

In his early forties a new obsession overtook Leonardo—mathematics—and his notebooks began to fill up with geometrical sketches and calculations.

LADISLAO RETI
The Unknown Leonardo

Preface

Leonardo da Vinci became unusually absorbed in geometry and remained so for a considerable length of time. Some writers have passed this off as an eccentricity of an extraordinary individual. This may have been. However, it would appear to be out of character for a person of such varied interests and talents. An inspection of Leonardo's notebooks shows that he did indeed occupy himself for extensive periods with geometrical endeavors. This same inspection, though, reveals that these efforts had a rather specific focus. Most of Leonardo's geometric sketches and diagrams are directly related to squaring curvilinear regions. Moreover, his methods are intuitive rather than algorithmic.

A study of just a few of Leonardo's insightful attacks quickly instills an appreciation of the power of dynamic dissection. It is easy to envisage this fifteenth-century Italian amazed by the clarity of transforming completely curvilinear regions into rectilinear regions of the same area. Certainly, this experience led him to seek other curved regions that could be rectified by similar commonsense methods. The highly satisfying success that he enjoyed in this endeavor served as a stimulus for further investigations. Moreover, he may have reasoned that if one became adept at squaring a variety of curved regions, the information and skill thereby gained might contribute to the goal that had eluded great minds for centuries: squaring the circle. Squaring the circle would bring unsurpassed recognition and glory to one who could finally achieve the goal sought by so many for so very long.

We now know that squaring the circle is impossible; there is

iii

evidence that Leonardo himself eventually became aware of this fact. But he still engaged in finding curvilinear figures that could be rectified. This activity was intrinsically rewarding to him.

Thus it appears that the effort Leonardo spent on geometry was *not* the result of his unusual makeup but rather the enticing nature of this particular geometric topic. He himself recognized this. Indeed, his notebooks reveal that he was preparing much of this material for a book he intended to write. This book was to be called "De Ludo Geometrico," which means something like "Fun with Geometry."

Perhaps as you explore the topic in the pages that follow you will find yourself catching Leonardo's excitement. This booklet was prepared to enable you, whether student or teacher, to investigate this part of geometry individually or with others—and to give you a taste for Leonardo's dessert.

Leonardo's "self-portrait," a rendition by Philip John Lange from *California MathematiCs* (November 1983). The drawing is thought by some to be Leonardo's depiction of his father.

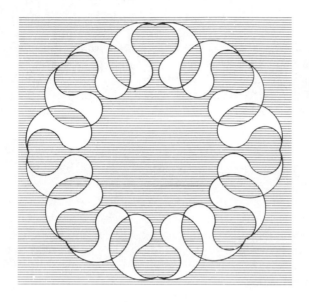

THIS fascinating figure, created by Leonardo da Vinci, is one of many designs that reveal this fifteenth-century genius's love of geometry. This particular one possesses several interesting features. For example, it consists entirely of semicircles, and these semicircles entail radii of only two different lengths. Further, the design illustrates Leonardo's success at placing several congruent shapes in a ring so that they fit together exactly. Symmetries abound. Later, we shall return to this delightful drawing and examine still another of its characteristics—one that will prove to be especially surprising.

Leonardo da Vinci loved geometry. Clearly, it was useful in his work as an artist and an engineer. Beyond that, however, geometry brought him personal challenge and enjoyment. His satisfaction from geometric endeavors is evidenced by this entry in his personal journal:

1509, April 28

Having for a long time sought to square the angle of two curved sides, that is the angle *e,* which has two curved sides of equal curve, that is curve created by the same circle: now in the year 1509, on the eve of the calends of May, I have solved the proposition at ten o'clock on the evening of Sunday. (MacCurdy 1941, 1160)

1

Others have commented on Leonardo's preoccupation with geometry:

> While there is no proof that he continued to study Latin, geometry became a real passion for the rest of his life. (Reti 1974, 72)

> In 1505 he composes Codex Forster I, which is dedicated to the transformation of one area into another or one solid into another equivalent one; and he fills up part of Codex Madrid II and hundreds of folios of the Codex Atlanticus with studies of the transformation of curvilinear surfaces into an equal number of rectilinear ones. (Marinoni 1980, 96)

> In his early forties a new obsession overtook Leonardo—mathematics—and his notebooks began to fill up with geometrical sketches and calculations. (Reti 1974, 67)

> His geometromaniac [sic] desire at this time to discover the more intricate relationships of area between circles, triangles, squares, polygons, segments, sectors, falcated triangles, and lunulae became an intellectual itch which he found impossible to scratch satisfactorily. Each new bout of scratching only served to stimulate fresh itches. (Kemp 1981, 296)

Thus, it can be said that geometry served Leonardo as an intellectual dessert. It was a subject he relished for its own particular taste and not for the utilitarian nourishment it may have provided. And whereas the measurement of curved regions generally requires complicated calculations involving the elusive pi, Leonardo's required only common sense—no pi at all.

Leonardo was especially fascinated by curvilinear regions that could be *rectified.* To rectify a curvilinear figure, one constructs a rectangle with the same area. For example, consider the pendulum in figure 1. Note that it is constructed entirely of *arcs.*

Fig. 1. The pendulum

Even though the boundary of the pendulum is completely curvilinear, its rectification is quite simple. We first inscribe the pendulum in a square (fig. 2a). We then rotate those portions of the figure that lie above the lower semicircular region (fig. 2b). They slip neatly into the empty region under the lower arc (fig. 2c). We see in figure 2d that the area of the pendulum is exactly that of the rectangle!

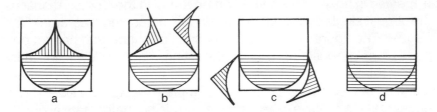

Fig. 2. Rectifying the pendulum

Now suppose the radius of each arc is 6. (Note in fig. 3 that the three arcs of the figure have equal radii. The radius of an arc of a circle is simply the radius of the circle of which the arc is a subset.) The diameter of the corresponding circle would be 12 and the area of the square surrounding the pendulum, 144. Hence, the area of the pendulum would be half of this, or 72. No pi! Even though our figure consists entirely of curves, our answer does not involve pi.

A second figure of the type that Leonardo investigated was the ax (fig. 4). The rectification of the ax is similar to that of the pendulum (fig. 5).

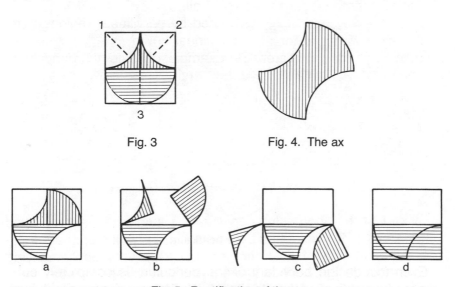

Fig. 3

Fig. 4. The ax

Fig. 5. Rectification of the ax

The Circular Figure

Let us now turn our attention again to the circular figure on page 1 (repeated in fig. 6). This figure can be found in the Codex Atlanticus, the largest single collection of Leonardo's notes and diagrams (1200 pages). The figure was cut out in a circular shape and pasted on a page of the manuscript. This is interesting, since so few folios in the Codex Atlanticus contain pasteups. Also pasted on the same page are four smaller fragments. The complete design, besides being unusual and attractive, possesses a surprising mathematical quality. Can you find the area of the whole unshaded region? Sure you can!

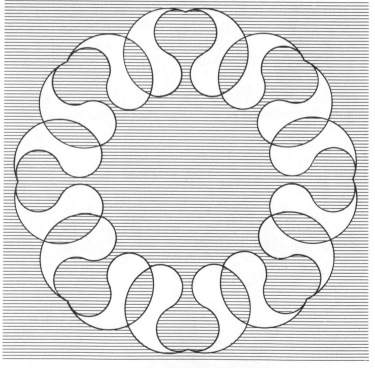

Fig. 6 One of Leonardo's intriguing figures

Note first that the design consists of a single element repeated eight times. Let's isolate that element (fig. 7), which appears independently in one of the four pasteup fragments accompanying Leonardo's design. Look at figure 8 and you will see that the entire motif comprises two large semicircles and four smaller ones.

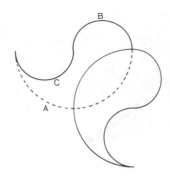

Fig. 7. Motif of Leonardo's circular design

Fig. 8. One of the motif's two large semicircles (A) is indicated with a dashed arc; two of the four smaller ones are marked B and C.

If we can find the area of this motif, multiplying it by 8 will give us the total area of the unshaded region in figure 6. Amazingly, we can find the area of the motif very easily because it, too, can be rectified. Study figure 9 to see how.

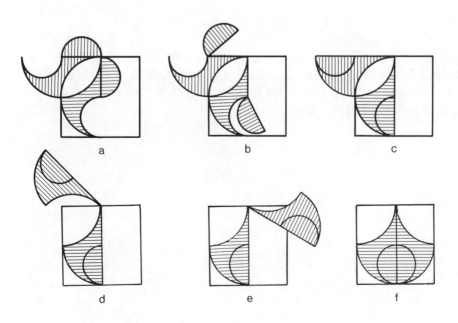

Fig. 9. Rectifying the motif of Leonardo's design

Now, let us suppose that the diameter of each of the larger semicircles is 4. What will be the length of a side of the square? What will the area be? Since the motif's area is that of the pendulum, it must be 8. Recall that there are eight motifs. Therefore, the area of Leonardo's complete design is 64.

The Crescent of Hippocrates

According to written records, the first person to rectify a curvilinear figure was Hippocrates of Chios (not the same Hippocrates who was the father of medicine). Hippocrates of Chios (ca. 450 B.C.) is credited with writing the first geometry text a full century before Euclid wrote his *Elements.* The curvilinear figure that he rectified consisted only of two arcs. It has the shape of a crescent (fig. 10).

Fig. 10. Crescent

Perhaps the simplest way to construct a crescent of Hippocrates is to draw semicircles on the sides of a square (fig. 11). Then circumscribing a circle about the square results in four crescents outside the circle (fig. 11d).

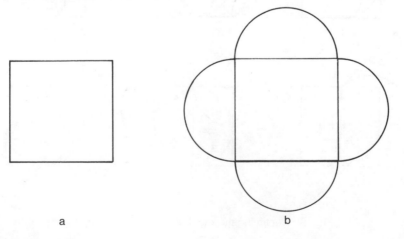

a   b

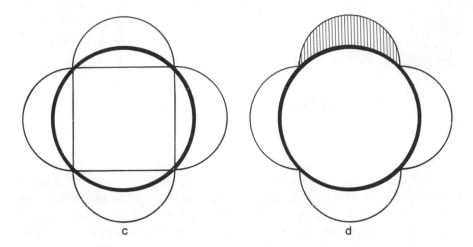

Fig. 11. Constructing a crescent of Hippocrates

Leonardo had a nice way of squaring these crescents, or *lunes*. The method was based on the fact that a circle circumscribed about a square has twice the area of a circle inscribed *within* the square. Here is a translation of Leonardo's own words:

> Circles made upon the same center will be double the one of the other if the square that is interposed between them is in contact with each of them. [See fig. 12.] And double the one of the other will be the squares formed upon the same center when the circle that is set in between them touches both the squares. [See fig. 13.] (MacCurdy 1941, 618)

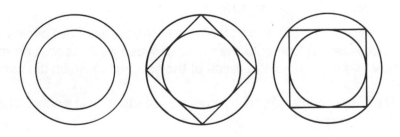

Fig. 12

7

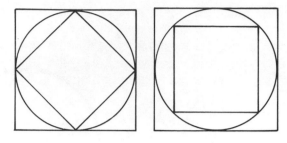

Fig. 13

The latter fact is proved, he explains, since the small square contains four of the eight triangles that make up the larger square. Figure 14c illustrates this nicely: The large square has been marked into eight congruent triangles, and since the small square contains four of them, it is evident that the smaller is exactly half the size of the larger.

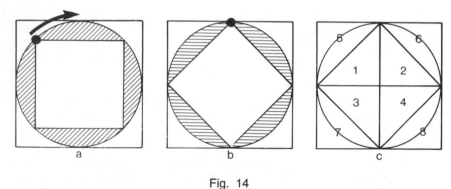

Fig. 14

To continue with Leonardo's explanation,

> There is the same proportion between circle and circle that there is between square and square, formed by the multiplication of their diameters. (MacCurdy 1941, 618)

He is saying here that to find the ratio of the areas of two circles it is not necessary to compute their areas or use pi. Instead, one may simply find the ratio of the areas of the squares in which the circles are inscribed.

At a different time he conveyed all this information in a succinct and charming way:

> When two circles touch the same square at four points, one is double the other. And also when two squares touch the same circle at four

8

points, one is double the other. [See figs. 12 and 13.] (MacCurdy 1941, 617)

From this last fact, Leonardo knew that the large circle in the four-lune construction (fig. 11c) had an area twice that of the circle on a side of the square (such as the shaded circle in fig. 15).

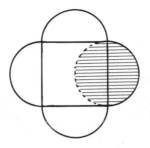

Fig. 15

Two of these small circles have the same area as four semicircular regions. So the figure obtained by removing the large circle from our complete figure (fig. 16b) has the same area as that resulting from taking off the four semicircles (fig. 16c). Thus, the four lunes (fig. 16b) have the same total area as the square. From this it follows that the area of a single lune is that of a square that is a fourth of the shaded square.

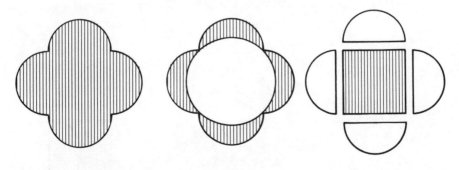

Fig. 16

An alternative approach allows the actual removal of the two smaller circles. Look at figure 17. Remember that removing the big circle leaves us with our four crescents. However, if we remove two of the smaller circles instead of the big one (fig. 17c), we will be left with

9

a figure having the same area as the crescents (fig. 17d). That figure consists of two congruent pendulums (fig. 18). Since each pendulum has an area that is half that of a surrounding square, their area together is that of the complete square.

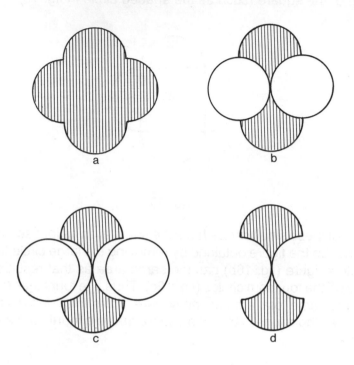

Fig. 17

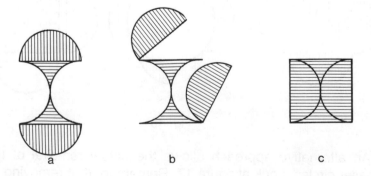

Fig. 18

10

Squaring the Cat's Eye

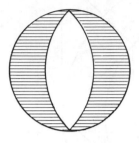

Fig. 19. The cat's-eye

The word for this squaring process in which we have been involved is *quadrature*. Let us now look at a different figure, the cat's-eye (fig. 19), and examine its quadrature.

The quadrature of the cat's-eye is important, since it is a simple application of a fact that Leonardo used extensively in his more complex rectifications of curvilinear regions. This fact concerns the segments of a circle that lie on the sides of an inscribed square (fig. 20) and the ratio of their areas to those of similar segments that lie within that square when another circle and another square are inscribed (fig. 21). Thus, the area of a segment of a circle that is circumscribed about a square has twice the area of a similar segment inscribed within the square.

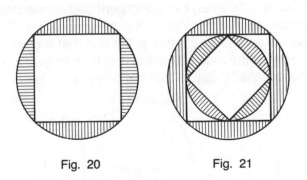

Fig. 20   Fig. 21

Leonardo carefully exhibited in a single figure many relationships based on this fact (see fig. 22). This figure shows the relations among segments of circles whose areas are in the ratio of a power of 2. This figure is typical of many such "sequence" diagrams that are found throughout Leonardo's notebooks. These diagrams tell complete stories in themselves without the need of discourse. This is especially fortunate for those not versed in Italian.

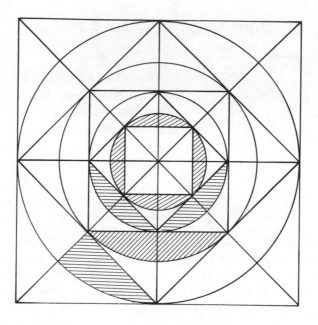

Fig. 22

Another basic component of many of Leonardo's designs is the *vesica*. A vesica results when two congruent circular segments that share a common chord are joined along that chord (fig. 23). Since vesicae consist of segments of circles, we find that the ratio of their areas is the same as the ratio of the areas of their respective circular segments. Indeed, the "pupil" of our cat's-eye is a vesica.

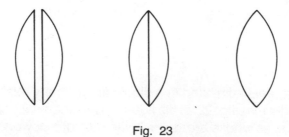

Fig. 23

Let us now look at the rectification of a cat's-eye (fig. 24). The total area of the four small segments of the circle that are outside the square (segments 1, 2, 3, and 4) is the same as the combined areas

12

of the two larger segments (5 and 6). Thus we can use the space occupied by the small segments to fill in the missing space at the pupil of the cat's-eye. This leaves us with the square inscribed in the circle (fig. 25).

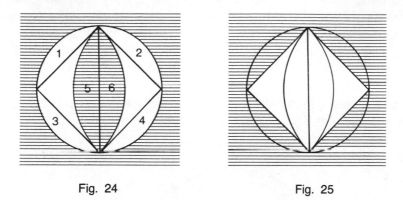

Fig. 24   Fig. 25

The alert reader has already noticed that a cat's-eye consists of two crescents of Hippocrates!

Figure 26 shows some figures that can be rectified using the relationships between segments and vesicae. You might like to try your hand at them.

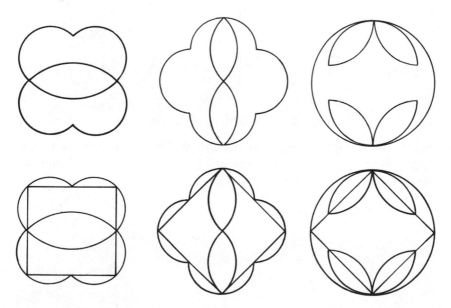

Fig. 26. Figures that can be rectified using the relationships between circular segments and vesicae

The Falcata

Leonardo made use of still another figure directly associated with segments of circles. This figure is a special kind of *falcata*. A falcata is a three-sided figure that has two curved sides and one straight side (fig. 27). Some dictionaries tell us that a falcata is a figure shaped like the blade of a scythe. Today, this is about as helpful as telling a student that an arbelos is a figure shaped like a shoemaker's knife. I've yet to see a shoemaker's knife. I should like to see one though, since it is supposed to look like an arbelos!

Fig. 27. A falcata

There are many falcatas of interest. If we cut a pendulum in half, we get two congruent falcatas (fig. 28). Indeed, Leonardo's entry in his personal journal quoted on page 1 refers to a falcata.

Fig. 28. 1 pendulum = 2 falcatas

Inscribing a circle within the design in figure 29 gives us still different falcatas that are easily squared (fig. 30). Our new circle has an area twice that of one of the smaller circles.

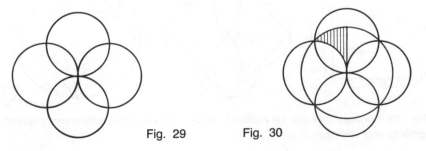

Fig. 29   Fig. 30

Would you like to see how the falcata is squared? Examine figure 31, following steps a–f. In step c we trade half of a circular segment for a full segment with the same area. (This equivalence is depicted in figure 22.) The full segment fits neatly under the remaining portion of the falcata, forming a triangle (d). The triangle is cut in half and flipped (step e), resulting in a square.

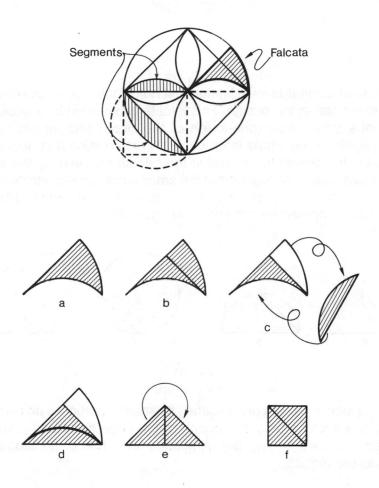

Fig. 31. Squaring the falcata

Just why Leonardo focused on falcatas that were half of a fully curvilinear figure is a mystery. For example, all three "sides" of the upper shaded portion of figure 32 are curves. Why didn't he work with such completely curvilinear figures?

15

Fig. 32

Instead, Leonardo investigated just half of this figure, as shown in the lower part of the design. This falcata is bounded by a *quadrant arc* of a small circle (one-fourth of the circle) and an arc that is one-eighth of the whole large circle. An interesting fact about this falcata also proves to be useful: the falcata has exactly the same area as a *quadrant segment* of the small circle (one-fourth its entire area). This can be established in several ways. Four different geometric approaches are shown in figure 33.

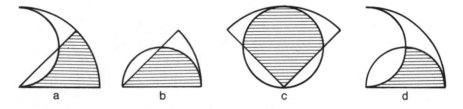

Fig. 33

Let's look at the first one in detail. Its shaded portion is an overlapping of a sector that is one-eighth of the large circle and a falcata called a *shark's fin* (fig. 34). A shark's fin is half of a Cleopatra's headdress (fig. 35).

Fig. 34. Shark's fin falcata

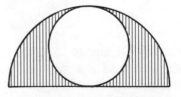

Fig. 35. Cleopatra's headdress

16

Surprisingly, the area of a Cleopatra's headdress is exactly that of the small inscribed circle. This can be verified by algebraic computation. However, it can also be shown geometrically. Consider figure 36. The diameter of the large circle is twice that of each of the small circles. So, the large circle has an area of four small circles. Since two circles have been "punched" out, the shaded region has an area of two small circles. Half of this region, the headdress of Cleopatra, will have the area of a single small circle. Furthermore, half of the headdress will have the same area as half of the small circle (fig. 37).

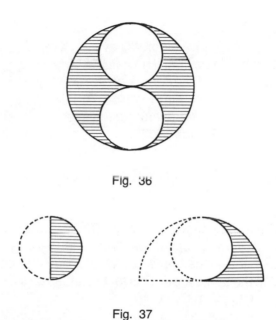

Fig. 36

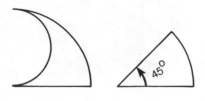

Fig. 37

Since a small circle has one-fourth the area of the large circle, half of the region of the small circle has one-eighth the area of the large circle. Consequently, the shark's fin has the same area as a sector that is one-eighth of the large circle: that is, a sector with a 45-degree central angle (fig. 38).

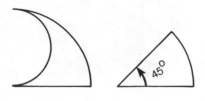

Fig. 38

Looking again at our overlapping figure (fig. 39), we see that removing the shaded figure (*S*) from the sector leaves us with the segment (*A*). However, if we remove the exact same shaded region from the shark's fin, we are left with the falcata (*B*). So, the falcata has an area equal to that of the segment. This reasoning can be expressed algebraically like this:

If $A + S = B + S$, then $A = B$.

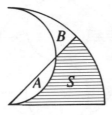

Fig. 39

All the remaining figures in figure 33 illustrate the same fact in different ways. In each case, removing the identical shaded region from equivalent regions leaves us with an unshaded falcata and a segment with the same area. This fact enables us to perform quadratures on the designs in figure 40.

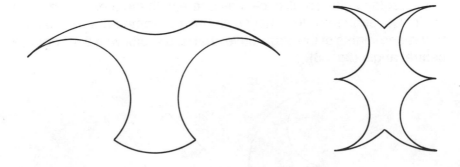

Fig. 40

18

Here's how (fig. 41).

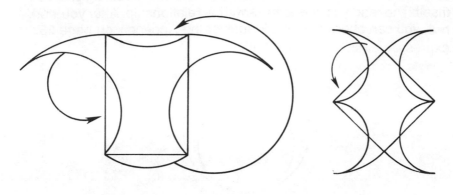

Fig. 41

The Claw

We'll conclude this introduction to Leonardo's quadratures by investigating his "claw" figures and some of the author's original figures based on them. Here is a pair of Leonardo's claws (fig. 42) and a picture of the foot of a bald eagle to help you with the imagery.

Foot of bald eagle

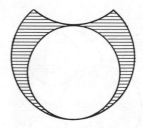

Fig. 42. A pair of Leonardo's claws

There are several ways to rectify this figure. You may like to give it a try yourself. To do so, however, you must know how it is made. The missing vesica, *m,* (fig. 43) consists of two quadrant arcs of the large circle. The area of the small circle is half that of the large one. This last fact is a consequence of the first one. It is an interesting exercise in itself. The reader is invited to verify this relationship. After you work it out, you can check your work against the Appendix on page 28.

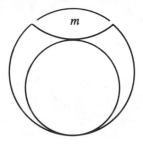

Fig. 43

We can proceed in a way similar to that of rectifying a lune of Hippocrates. That is, we shall rectify a single pair of claws by rectifying four pairs of claws. Consider the design in figure 44. We can find the area of the four pairs of claws by using the "haystack principle." It is well known that it is hard to find a needle in a haystack. One cunning way to do so is to burn down the haystack. What's left is the needle. Here our haystack is the whole interior of the outer boundary (fig. 45). If we burn away one bigger circle (fig. 46a) and then burn off four smaller ones (fig. 46b), we are left with our desired needle—the claws.

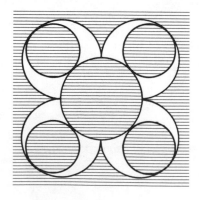

Fig. 44   Fig. 45

20

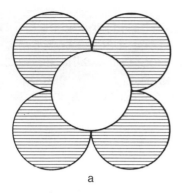
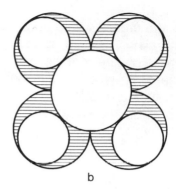

a   b

Fig. 46

To look at the design another way, the whole bounded figure consists of four big circles and one concave diamond (fig. 47).

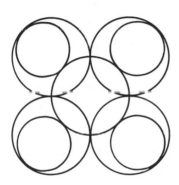
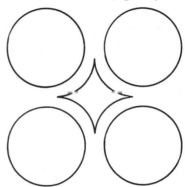

Fig. 47

With this in mind, we can square the claws in an elegant manner. The circles involved in the design have areas in the ratio of 2:1. So instead of burning off one big circle and four small ones, we can remove three of the big ones (fig. 48).

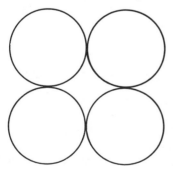

Fig. 48

21

We are left with a new "needle," an urn having the same area as the claws. The urn is easily squared. Notice that the concave diamond (the neck of the urn) is made up of four corners that lie outside the circle when it is inscribed in a square (fig. 49). Although the corners are touching, they can be rearranged around the circle to complete the square.

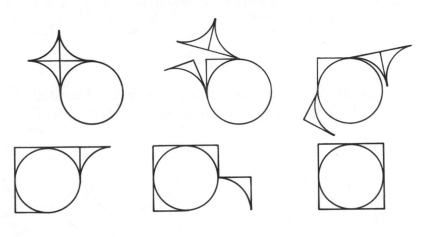

Fig. 49

Thus, four pairs of claws have an area equal to that of a square that circumscribes one of the large circles.

Summarizing the quadrature

Let us now look at a summary of the complete quadrature of four pairs of claws. We shall need to recall the makeup of the whole figure (which we shall refer to in our equation as WHOLE) that yielded the four pairs of claws (CLAWS) at the outset:

$$(*) \quad \text{WHOLE} = 4 \text{ BIG CIRCLES} + 4 \text{ CORNERS}$$

This fact, $(*)$, will be helpful in following the outline of the quadrature of claws exhibited here:

```
CLAWS = WHOLE - 1 BIG CIRCLE - 4 SMALL CIRCLES
      = WHOLE - 1 BIG CIRCLE - 2 BIG CIRCLES
      = WHOLE - 3 BIG CIRCLES
      =              1 BIG CIRCLE + 4 CORNERS      using (*)
      = A SQUARE SURROUNDING A BIG CIRCLE
Hence, CLAWS = A SQUARE SURROUNDING A BIG CIRCLE.
```

22

Leonardo's method

Leonardo had a more direct approach to this result. His method relied on the ratio of segment areas.

Let's look at the figures in the left column of figure 50, the crescent and the small circle below it. Since the whole large circle of the crescent has twice the area of the missing small circle, the area of the crescent is exactly that of the small circle below it. Now, if we subtract equivalent areas from these equal figures—the crescent and the small circle—we will be left with regions having equal areas. This is

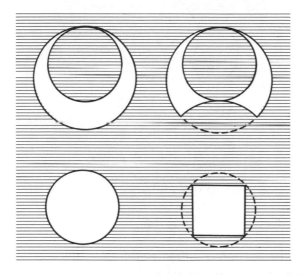

Fig. 50

shown in the right column. From the crescent we have removed a vesica. From the small circle we have removed four segments whose total area is that of the vesica (as you will remember from the cat's-eye problem earlier). Thus, the area of the pair of claws is the same as that of the square inscribed in the small circle. This is summarized by a single illustration (fig. 51).

Such claws can be used as elements in original designs. For example, reversing one of the claws from each pair gives us this attractive design (fig. 52). Not only is it eye catching, but it is also rectifiable.

23

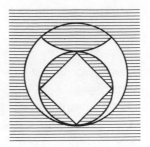

Fig. 51

Fig. 52

Figure 53 shows some other rectifiable designs that involve Leonardo's claws.

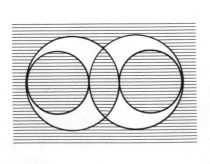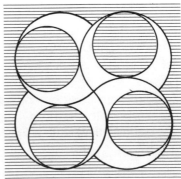

Fig. 53

We can also use combinations of the figures discussed to provide opportunity for further creativity and problem solving, for instance, those in figure 54.

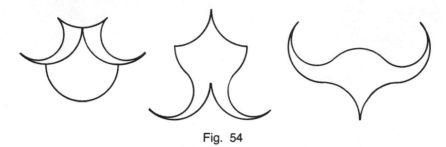

Fig. 54

Figure 55 shows some illustrations that result from inscribing axes or pendulums within pairs of claws.

Fig. 55

Of course, it is not necessary to be restrictive about where one deletes the vesicae. We may remove them as shown in figure 56 to get a more pleasing catlike figure.

Fig. 56

Some additional figures based on the principles discussed appear in figure 57.

Figures 40, 41, 52, 53b, 54, and 56 are original designs of the author.

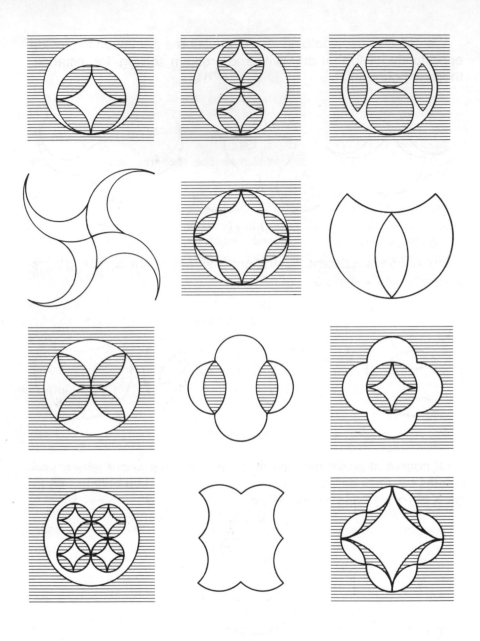

Fig. 57

You may wish to try your hand at rectifying them. If so, then we suspect that you have already begun to acquire a taste for Leonardo's dessert.

BIBLIOGRAPHY

Clark, Kenneth, and Carlo Pedretti. *The Drawings of Leonardo da Vinci in the Collection of Her Majesty the Queen at Windsor Castle.* London: Phaidon, 1968.

Gwilt, Joseph. *The Encyclopedia of Architecture: Historical, Theoretical, and Practical.* New York: Crown Publishers, 1982.

Heath, Thomas. *A Manual of Greek Mathematics.* New York: Dover Publications, 1963.

Hobson, Ernest. "Squaring the Circle." In *Squaring the Circle and Other Monographs.* 4 vols. in 1. New York: Chelsea Publishing Co., 1969.

Kemp, Martin. *Leonardo da Vinci: The Marvelous Works of Nature and Man.* Cambridge, Mass.: Harvard University Press, 1981.

Lawlor, Robert. *Sacred Geometry: Philosophy and Practice.* The Illustrated Library of Sacred Imagination, edited by Jill Purce. New York: Cross-Road Publishing Co., 1982.

Leonardo da Vinci. *The Codex Atlanticus of Leonardo da Vinci* [facsimile of the collection]. 12 vols. Florence: C. E. Giunti-Barbera, 1975.

Leonardo da Vinci. *Codex Madrid II* [facsimile of the collection]. New York: McGraw-Hill, 1974.

Leonardo da Vinci. *Codice Atlantico di Leonardo da Vinci* [facsimile of the collection]. Milan: Ulrico Hoepli, 1939.

MacCurdy, Edward, ed. *The Notebooks of Leonardo da Vinci.* New York: Garden City Publishing Co., 1941.

Marcolongo, Roberto. *Indici del Codice Atlantico di Leonardo da Vinci.* Milan: Ulrico Hoepli, 1939.

Marcolongo, Roberto. *Leonardo da Vinci: Artista - Scienziato.* Milan: Ulrico Hoepli, 1950.

Marinoni, Augusto. "Leonardo's Writings." in *Leonardo the Scientist,* edited by Ladislao Reti. New York: McGraw-Hill Book Co., 1980.

Pedoe, Dan. *Geometry and the Liberal Arts.* New York: St. Martin's Press, 1976.

Pedretti, Carlo. *Leonardo da Vinci Fragments at Windsor Castle.* London: Phaidon, 1957.

Reti, Ladislao. *The Unknown Leonardo.* New York: McGraw-Hill, 1974.

Vallentin, Antonina. *Leonardo da Vinci: The Tragic Pursuit of Perfection.* Translated by E. W. Dickes. New York: Viking Press, 1938.

Wills, Herbert. "Leonardo da Vinci's Design." *School Science and Mathematics* 83 (October 1983): 453–63.

———. "Socrates, Hippocrates, Leonardo da Vinci and Some Interesting Areas." California MathematiCs 8 (November 1983).

Zammattio, Carlo, et al. *Leonardo the Scientist.* New York: McGraw-Hill, 1980.

Appendix

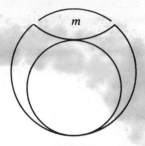

Fig. 43

The problem

Show that the area of the small circle in figure 43 is half that of the large one, given that the missing vesica, *m*, consists of two quadrant arcs of the large circle.

The solution

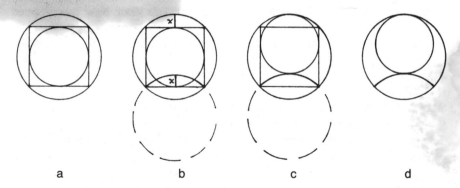

Fig. 58

1. Figure 58a shows our old friend the square with a circle inscribed in it and another circle circumscribed about it. Recall that in such a situation the inner circle's area is half that of the outer circle (p. 7).
2. Figure 58b shows two congruent quadrant segments each having a maximum height of x.
3. Now move the inscribed circle upward a distance of x (fig. 58c).
4. Delete the square (fig. 58d). Clearly, the small circle in the figure is congruent to the small inscribed circle of figure 58a. Thus we can see that the area of the small circle is half that of the larger circle.